A Theoretical Model of Feedback in Pharmacology Using Signal Flow Diagrams

By

Bradley Scott Tice

ISBN: 0-75962-644-8

This book is printed on acid free paper.

1stBooks - rev. 7/24/01

Submitted to Fairfax University, USA as
a part of the Learning Plan aspect of the
programme requirements leading to the
degree of Doctor of Philosophy
Chemistry (Ph.D.).
August 1996

Abstract

The dissertation will model a theoretical system of pharmacological drug feedback by the use of signal flow diagrams. A comprehensive evaluation of a signal flow diagram verses a block diagram and the importance of the signal flow diagram will be examined. A clinical study using specific drugs and specific problems, i.e. immunological functions, on human and animal test subjects from actual case studies will be used to model the feedback system with the signal flow diagrams. The results from this study will present the signal flow diagram as a clear, accurate and easy method of pharmacological feedback.

Acknowledgement

I would like to dedicate this dissertation to my Father and Mother who have made it all possible and to an old friend Mr. Sam Zambre who has been a scout master, employer and friend to me for over twenty-five years. I would also like to thank all the people over the years, who are too numerous to list, who allowed me to explore, challenge and grow as a person and I hope that someday soon I can repay this great debt.

The trouble with chemists is that chemistry is too difficult for them.

Albert Einstein

Contents

Introduction

The dissertation will model a theoretical system of pharmacological drug feedback by the use of signal flow diagrams. A comprehensive evaluation of a signal flow diagram and then a comparison with traditional formal or block diagrams will be examined. A clinical study using specific drugs and specific problems, immunological studies taken from actual clinical case studies, will be used to model the feedback system with the signal flow diagrams.

They will be in three separate segments, Model A, B, and C of which Model A will discuss delayed hypersensitivity skin testing. Model B will be the triple-test plan for serologic diagnosis of syphilis and Model C the laboratory effects of cyclosporine. The results from this study will present the signal flow diagram as a clear, accurate and easy method of pharmacological feedback.

A practical application can be made for using signal flow diagrams to graph pharmacology feedback by the health professions along the lines of medical flow charts that are used to reduce errors, burnout and redundancy in patient care (Cronin, Lane and Peirce, 1983).

Literature Survey

Feedback as defined by the engineering concepts of control systems started in the 1920's and 1930's and was first given voice in Minorsky's paper on the steering of ships in 1922 and followed by Nyquists' paper on Regeneration Theory, 1932, and Hazen's paper on the Theory of Servomechanisms in 1934 (Chang, 1961: 1). Brown and Whiteley's work on servomechanism theory and Tustin's work on non-linear elements in servo-design are also worth mentioning (Porter, 1950: 2). The first practical application of feedback control was James Watt's flyball governor to the steam engine in 1775 (Murrill, 1967: 4).

A system is defined as an orderly combination or arrangement of parts into a whole and that these combination of parts represent a methodological arrangement that represent a system (Koenig and Blackwell, 1961: 1). A feedback control system is a combination of elements which cooperate to maintain a physical quantity, the output, equal to the desired output that is related to other physical quantities known as inputs (Newton and Gould, 1957: 1-2). A feedback system is distinguished from a network by the presence of at least one unilateral element that represents power, information, or a commodity that can flow in only one direction (Smith, 1958: 1).

All automatic regulating systems can be divided into two groups, the direct-acting and the indirect-acting systems. The direct-acting systems is were the action of the sensitive element on the regulatory organ is operated with the introduction of additional source of energy and the indirect system is when the sensitive element acts on the regulatory organ directly through a special amplifier as an auxillary source of energy (Popov, 1962: 17-19). Feedback can be considered a form of communication in which an input is responded to by an output (James, Nichols and Phillips, 1964: 1). The loss of time between actions and reactions is the major point of control systems problems and represents the measure of a disturbance (Holzbock, 1958: 3).

A closed-cycle control system is any signal that is supplied to the controller as a function of the object or device being controlled (Zeines, 1959: 22), (Horowitz, 1963: 58) and (Ku, 1962: 10). A physical system is stable if, when disturbed from its equilibrium state, it can ultimately return to that original condition, it is unstable if it increases indefinitely with time (Hardie, 1964: 188). The parameters of such a system must be 'optimally adjusted' to insure optimum control action and is the heart of the feedback system (Oldenbourg and Sartorius, 1953) and (Kipiniak, 1961:1).

The selection of these primary measuring elements are important in that they must conform to the control requirements of the process in controlling the variables (Hadley and Longobardo, 1963: 4). Most control feedback systems are defined as servomechanisms and represent a closed-loop system (Chubb, 1967: 3) and (Gille, Pelegrin and Decauline, 1959: 15). The elements which constitute the essentials of a servomechanism are also the inherent properties of a vibrating system and can be detrimental to a servomechanism (Bulliet, 1967: 92). A power control system is considered to be an object of automation and servo-systems the tools to automation (Bulgakov, 1965: vii).

The sole purpose of a control system is to minimize the 'process-upset' when such a disturbance arises (Tucker and Willis, 1958: 3). The use of 'set points' as an ideal or specified process are the guidelines used to keep the control in a system and are used to measure all deviant actions to maintain control (Rusinoff, 1957: 19). Modelling of such a system allows for the evaluation of systems that would otherwise be beyond the direct observation of a feedback system but special care must be taken to insure that the model accurately describes the function it is modelling (Seifert and Steeg, 1960: 15) and (Derusso, Roy and Close, 1965:1).

Feedback systems are all dynamic systems and fundamentally similar and operate on the difference between the actual state of the system and the arbitrarily varied ideal state (Ahrendt, 1954: 1) and (Bowen and Schultheises, 1961: 4). Feedback is also a form of self-regulation and are an inherent process in both living and non-living processes of nature (Nagel, 1948: 2). In defining sub-units of a feedback system, elementary components should be defined by there functions rather than by appearance and that before a physical quantity can be controlled, it is essential that it be measurable (Warren, 1967: 9), (West, 1953: 12) and (Qvarnstrom, Schurt, and Runnstrom-Reio, 1965: 41).

The mathematical measurement of feedback systems is usually the domain of differential equations in that such processes are based on a present state that is instantaneous rather than a past history of actions (Oguztoreli, 1966: 3). The use of linear differential equations assumes the linearity of the properties being formulated although such characteristics do not reflect the true nature of the real world as no system is completely linear or non-linear also the mathematical modeling of systems is difficult when working with 'optimizing' and 'adaptive' systems (Thaler and Pastel, 1962: 1) and (Peschon, 1965: 10).

A differential equation is linear if the equation contains only first powers of the dependent variable or its time derivatives (Tsien, 1954: 1). A non-linear equation is the result of higher powers of the variable and cross products of these variables and its derivatives (Tsien, 1954: 1). In designing an 'optimal' control system, four kinds of variables must be accounted for in the design. Independent variables which are the manipulating variables for the control or monitoring of the process. Dependent variables that serve to measure and describe the state of the process at any instant of time. Product variables that are used to indicate and measure the quality of the operating performance of the control system and fourth, the Disturbances which are the uncontrollable, environmental variables (Tou, 1964: 8).

The use of the Root Locus technique is used to determine time domain properties and is a graphical method of determining the roots of the characteristic equation of a single loop system (Kuo, 1962: 246). The use of Fourier and Laplace transformations are used to analyse transient processes in linear systems where signals appear as prescribed functions of time (Solodovnikov, 1965: 17). The two most important determining factors of the use of differential equations is what is the type of system and the manner in which the specifications are set forth as performance of the system can only be examined by the 'appropriate' solutions to these equations (Wilts, 1960: 9).

The mathematical definition of adaptability is not a precise one due to the fact that feedback has various definitions and permutations (Mishkin and Braun, 1961). Physical systems have many common characteristics that can be described by mathematical modeling and assigned definite groups according to the structure of these models (Tomovic, 1963: 4). Feedback adds to the complexity of a system and hence adds to the complexity and difficulty of the analysis of such systems (Lindorff, 1965).

Another term for feedback systems is cybernetics. The leading figure in the field of cybernetics is the mathematician Norbert Wiener and is best embodied in his famous book <u>Cybernetics</u> (Cambridge: MIT Press, 1948). Wiener states "Messages are themselves a form of pattern and organization." Indeed, it is possible to treat sets of messages as having an entropy like sets of states of the external world. Just as entropy is a measure of disorganization, the information carried by a set of messages is a measure of organization. In fact, it is possible to interpret the information carried by a message as essentially the negative of its entropy, and the negative logarithm of its probability. (Wiener, 1950).

Another term for feedback is homeostasis. Cybernetics is a Greek word for pilot or steerman. Homeostasis is the balancing of the bodies systems to produce and support life. Strict limits, especially when dealing with higher life forms, is necessary in defining the conditions that make up the properties of homeostasis and tend to operate more slowly than the voluntary or postural feedbacks (Wiener, 1948: 114). William Cannon's seminal work <u>The Wisdom of the Body</u> is a prime example of homeostasis in the human body and cites clear examples of these processes operating in the organs and systems of man (Cannon, 1932).

The nervous system is a system which controls the interrelationship between the organism and the environment and regulates the internal environment, the process of homeostasis (Adolph, 1960: 13). Although Cannon was the first to use the term homeostasis, he paid homage to the German physiologist Pfluger and the Belgian physiologist Fredericq and the ancient Hippocrates for the generation of this concept (Langley, 1965: 1).

Cannon considered homeostasis as a defense protection of the body by the preservation of consistency in the fluid matrix of the body against unfavorable conditions that would arise if not preserved (Cannon, 1932: 202). Homeostasis is derived from the term 'homeo' which means like or similar, and 'stasis' meaning standing. Metabolism has also been applied to the 'inflow' and 'outflow' of matter and energy and intermediary transformations within the organism (Henderson, 1913: 24).

Another definition of feedback is communication theory and information science. Lwoff has defined information as, following Brillouin's analysis, if considering a certain number of possible answers, when no information is avaliable, and when some information is gained, then the number of possible answers is reduced, and the complete information means only one answer, in that information is a function of the ratio of possible

11

answers 'before and after' (Lwoff, 1962: 91). Communication theory is best represented by Shannon and Weavers seminal work <u>The Mathematical Theory of Communication</u> (Urbana: University of Illinois Press, 1949) that, like Norbert Wiener's <u>Cybernetics</u>, helped define early communication theory.

Information science is a general term used to cover the fields of library science, computer networks and cognitive and neural sciences. Types of feedback are represented in all of these disciplines. Two works that deal with information as a quantitative science are Edward R. Tufte's <u>The Visual Display of Quantitative Information</u> and <u>Envisioning Information</u> (Tufte, 1983 &1990).

The basic characteristics of a feedback system, as defined as an automatic-control system, can be illustrated qualitatively by the use of a block diagram (Thaler, 1955: 2). A block diagram indicates a signal transfer where a circuit diagram represents also the state of energy transfer (Izawa, 1963: 19). Feedback control systems function in the human body and represent some of the most complex systems in the world (Instrument Society of America, 1958: 5).

All types of feedback control systems can be described in terms of functional block diagrams. The blocks in a diagram are interpreted as representing functions of components and not as isolated pieces of equipment as several functions can be combined into a single piece of physical equipment or several pieces will be required to perform a single function (Doebelin, 1962: 5). The block diagram represents the operations performed in a system and in a manner in which the signal information flows throughout the system (Del Toro and Parker, 1960: 5). A block diagram is a graphical representation of interconnected elements or components which form a system and differ only in their dynamic properties (Solodov and Fuller, 1966: 44).

The signal flow diagram has the conventional form of a circuit diagram, comprising connected branches and nodes, and differs from the circuit diagram in two important ways. The nodes are the points where the signals appear and where they are summated, and that the branches are oriented and their orientation forms a single channel for the travel of the signal (Macmillian, 1964: 4- 5).

The following is a graphical representation, block diagrams, of a basic feedback system (West, 1960: 15).

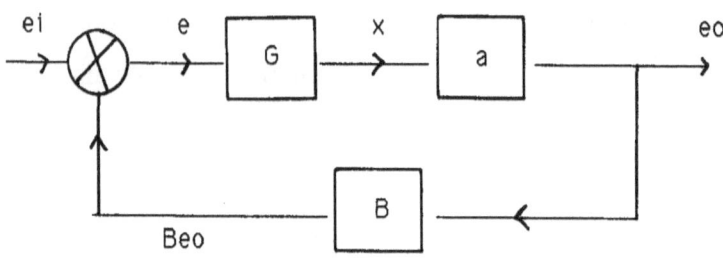

The original amplifier with gain 'a' has an input and output signal 'x' and an output e related by eo= a x. An additional amplifier of gain 'G' has an input and output signals 'e' and 'x' respectively with x= Ge. The signal e is feed back through a passive network B to form a feedback signal Beo. A comparator or error detector indicated by a circle is fed with an input signal e and the feedback signal Be to form the signal 'e' defined by e=ei-Beo. The whole system is expected to operate as an amplifier of gain 'a' with improved reliability (West, 1960: 15-16).

This is an example of a generic diagram of a adaptive control system.

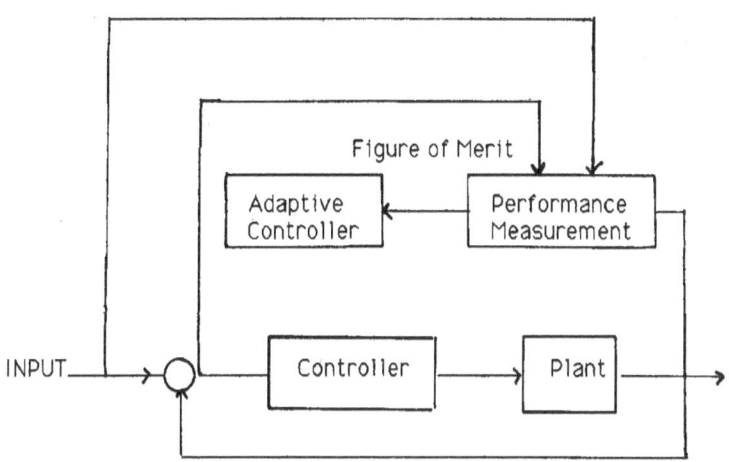

The adaptive control system has three necessary performance measurements (Caruthers and Levenstein, 1963: 1).

1. Means for a performance measurement.
2. Means for translating this into a quantitative figure of merit.

3. A closed-loop control of system parameters to achieve an optimum as given by the figure of merit.

The performance measurement is assumed to be based on observation of system input, output , and /or error signal. The figure of merit or error signal is derived from these through the definition of the performance criterion. The adaptive controller adjusts controller parameters to achieve an optimum (Caruthers and Levenstein, 1963: 2).

These two diagrams are to give a visual reference to feedback and adaptive control systems that have been mentioned in the course of this text.

In reviewing biomedical papers using feedback systems the following was found. The use of Bayesian networks for drug delivery optimization (Bellazzi, 1993). Biological modeling on a microcomputer using a standard spreadsheet and equation solver programs (Plouffe, 1992). Advanced computer programs for drug dosing that combine pharmacokinetic and symbolic modeling of patients

(Lenert, Lurie, Sheiner, Coleman, Klostermann and Blaschke, 1992).

Open-loop stochastic control of pharmacokinetic systems (Lago, 1992). Analysis of low-frequency lung impedance in rabbits with non-linear models (Peslin, Saunier, Duvivier and Marchand, 1995). An alternative pathway for signal flow from rod photoreceptors to ganglion cells in mammalian retina (DeVries and Baylor, 1995). Validation of continuous thermal measurement of cerebral blood flow by arterial pressure change (Wei, Shea, Saidel and Jones, 1993). Problems of selecting the reagent concentration in the solid-phase immunoenzyme method for determining antibody concentrations using block diagrams (Panteleev, Vaneeva, Demchenko and Semenova, 1987).

A new design concept for an audio dosimeter described by block diagrams (Clark, 1980). Conceptions for optimization in radiation therapy using block diagrams (Hubener, 1978). Contrast echocardiography in clinical practice (Dubourg, Chikli and Delorme, 1995). A new analysis method for disposition kinetics of enterohepatic circulation of diclofenac in rats (Fukuyama, Yamaoka, Ohata and Nakagawa, 1994). Discussion of the charging and discharging circuits and present the general block diagram of an automatically triggered current defibrillator (Monzon and Guillen, 1985). Differences between normal blood pressure regulation in the circulatory system under normal, or equifinal conditions are analyzed using hydraulic models, block diagrams and signal flow diagrams (Guenther, Morgado and Penna, 1974).

Testing fiber optic data links for sensitivity to high gamma radiation dose rates using block diagrams (Krinsky, 1988). A simple device for photographic dosimeter calibration using block diagrams (Ladu and Randaccio, 1980). A more simple PROM programmer using block diagrams (Coco, 1979). Measuring, controlling and dose-rating with miniature oval-gear counters using block diagrams (Feil, 1979). New developments in X-ray television using block diagrams (Heister, 1979). Modeling, testing, simulation, and optimization of digitalis pharmacokinetics using block diagrams (Ranjbaran, 1974). Models of functional shutdown of 1 hemisphere and neuro pharmacological effects on deep cerebral structures using block diagrams (Menshutkin, Suvorova and Balonov, 1981).

Television microscopes for studying biological microscopic material using block diagrams (Danilov, 1984). State variable techniques in pharmacokinetics using computer graphics and block diagrams (Ranjbaran, 1980). The papers dealing with graphical information of feedback almost exclusively use block diagrams for visual representation and this follows the long tradition of using such methods in the fields of engineering, information science and design.

Signal Flow Diagram

The use of signal flow diagrams are common in fields such as engineering and a practical use of them can be made in the field of pharmacology. The main reason for the use of signal flow diagrams over other diagram systems, formal or block diagrams, are that they are easy to use and permits a solution practically upon visual inspection (Shinners, 1964: 25)[1] . Signal flow diagrams can solve complex linear, multiloop systems in less time than either block diagrams or equations (Macmillian, Higgins, and Naslin, 1964: 4). A signal flow graph is a topological representation of a set of linear equations as represented by the following equation

$$\text{Equation 1:} \quad y_i = \sum_{j}^{n} a_{ij} x_j, \quad i = 1,\dots, n$$

---

[1] Original work for this dissertation was done in the form of an unpublished paper for Advanced Human Design in Cupertino, California U.S.A. A copy of the paper is in Appendix A.

Branches and nodes are used to represent a set of equations in a signal flow graph. Each node represents a variable in the system, like node i represents variable y in equation 1. Branches represent the different variables such as branch ij relates variable yi to yj where the branch originates at node i and terminates at node j in equation 1 (Shinners, 1964: 25).

The following set of linear equations are represented in the signal flow graph in Figure 1 (Shinners, 1964: 25).

$$y2 = \quad ay, \quad +by2 \qquad +cy4$$
$$y3 = \qquad \quad dy2$$
$$y4 = \quad ey1 \qquad +fy3$$
$$y5 = \qquad \qquad +gy3 \quad +hy4$$

Figure 1

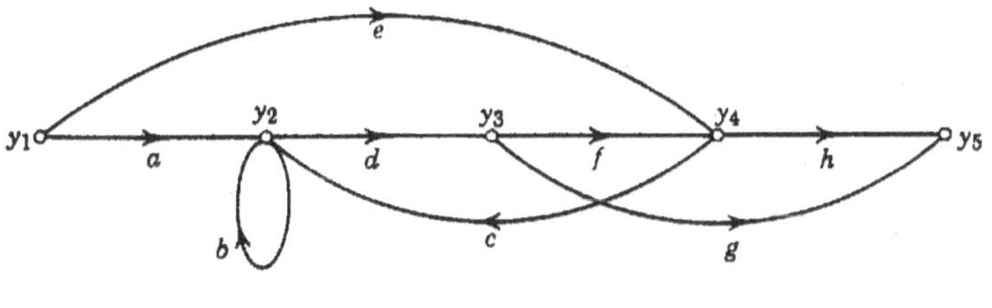

21

It is necessary now to define the terms as represented by the signal flow diagram in Figure 1 (Shinners, 1964: 28).

1. The <u>Source</u> is a node having only outgoing branches, as y1 in Figure 1.
2. The <u>Sink</u> is a node having only incoming branches, as y5 in Figure 1.
3. The <u>Path</u> is a group of connected branches having the same sense of direction. These are he, adfh, and b in Figure 1.
4. The <u>Forward Paths</u> are paths which originate from a source and terminate at a sink along which no node is encountered more than once, as are eh, adg, and adfh in Figure 1.
5. The <u>Path Gain</u> is the product of the coefficient associated with the branches along the path.
6. The <u>Feedback Loop</u> is a path originating from a node and terminating at the same node. In addition, a node cannot be encountered more than once. They are b and dfc in Figure 1.

7. The <u>Loop Gain</u> is the product of the coefficients associated with the branches forming a feedback loop.

By using a signal flow diagram to represent the variables associated with pharmacological testing, drug delivery, behavior, dosage and time intervals can all be graphed for ease of representation of these complex systems.

Signal Flow verses Block Diagrams

The strongest points to be made in favor of the signal flow diagram over block diagrams, sometimes also termed formal diagrams, is that they are more general in use, being applied in a more collective or universal manner, and that they are a more pronounced rationalization stemming from diagrammatic simplification (Macmillian, Higgins, and Naslin, 1964: 4)[2] .

A block or formal diagram is a graphical representation of interconnected elements which form a system that differ only in their dynamic properties (Solodov, 1966: 44). A signal flow diagram represents a set of equations by means of branches and nodes. The following are examples of how complex and visually daunting block diagrams can become with multiple equations (Shinners, 1964: 27).

---

[2] An unpublished paper for Advanced Human Design titled "Signal Flow Diagrams verses Block Diagrams: The Simpler Road" is the bases for this section and a copy of the paper can be found in Appendix B.

Figure A

Original System

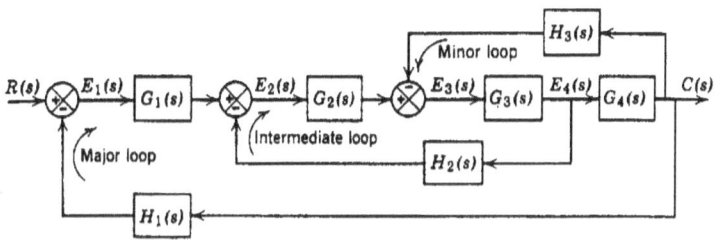

Figure B

Rearranged The Summing Points of The

Intermediate and Minor Loops

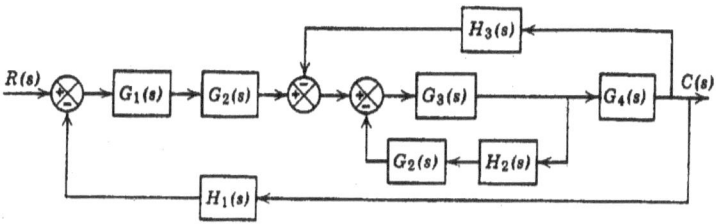

Bradley Tice

Figure C

Reduced The Equivalent Intermediate Loop

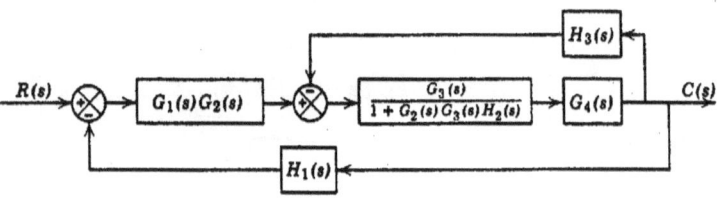

Figure D

Reduced The Equivalent Minor Loop

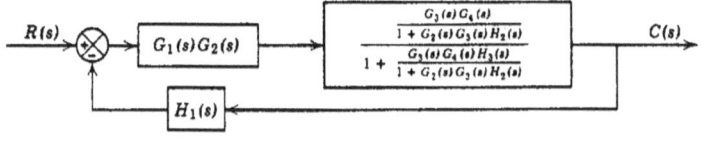

Figure E

The Equivalent Feedback System

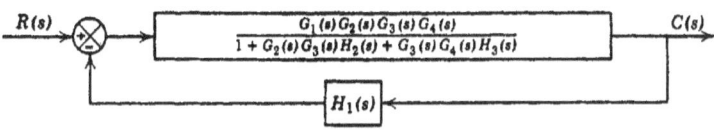

Figure F

The System Transfer Function

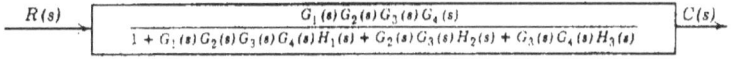

It becomes quite clear that such diagrams have many serious limitations of which the lack of ease, accuracy and efficiency are the major flaws that can all but be eliminated by the use of signal flow diagrams.

A Pharmacological Model

In applying a signal flow diagram to a pharmacological system of drug feedback it is important to have actual clinical data for source material. All of the data is from actual clinical and laboratory studies.

Each of the sections is divided into three test models. Model A that deals with delayed hypersensitivity skin testing. Model B that is the triple-test plan for serology diagnosis in syphilis and Model C on the drug cyclosporine. Information of each of these sections, i.e. drug descriptions, testing methods and testing standards, are found in the appendix.

In Model A the signal flow diagram graphs the rank by percentile of reactions to the six antigens of the hypersensitivity skin test. The sample size was 76 normal adult subjects and the following is the results of the six antigens, plus two agents that are also known indicators of possible anergy; Coccidioidin and Mumps, as represented in Table 1.

Table 1

Antigen	Intermediate Strength	Second Strength
Candidin	39%	92%
Coccidioidin	19%	45%
Mixed respiratory vaccine	41%	n/a
Mumps	78%	n/a
PPD	26%	83%
SK-SD	55%	93%
Staphage Lysate	71%	n/a
Trichophytin	28%	n/a

The rank of percentile can be indicated by the following signal flow diagram as represented in Figure 2.

Figure 2

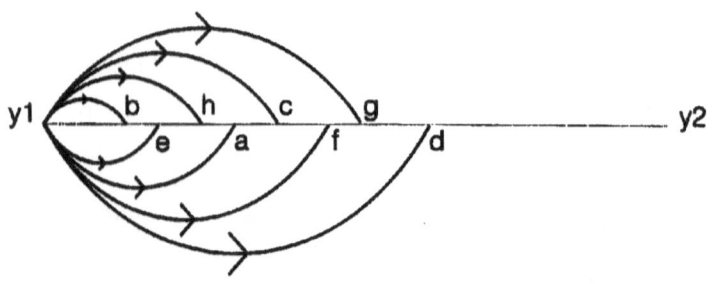

The following table, Table 2, represents the variables, i.e. antigens, as they rank in percentile[3] .

Table 2

y1 is the source.

a is Candidin at 39%

b is Coccidioidin at 19%

c is Mixed respiratory vaccine at 41%

d is Mumps at 78%

e is PPD at 26%

f is SK-SD at 55%

---

[3] For more information on hypersensitivity skin testing see Appendix C.

g is Staphage Lysate at 71%

h is Trichopytin at 28%

y2 is the sink.

From this table, Table 2, and figure, Figure 2, the best antigen for evaluating anergy is Staphage Lysate, at 71%, with SK-SD following in second with 55%. The 78% rating for Mumps is greater than Staphage Lysate but was not a part of the six antigens scheduled for use.

In Model B the Triple-Test Plan for Serologic Diagnosis of Syphilis is taken from a flow chart in <u>Clinical Laboratory Diagnosis</u> (Levinson and MacFate, 1969: 672) and is represented by Figure 3.

Figure 3

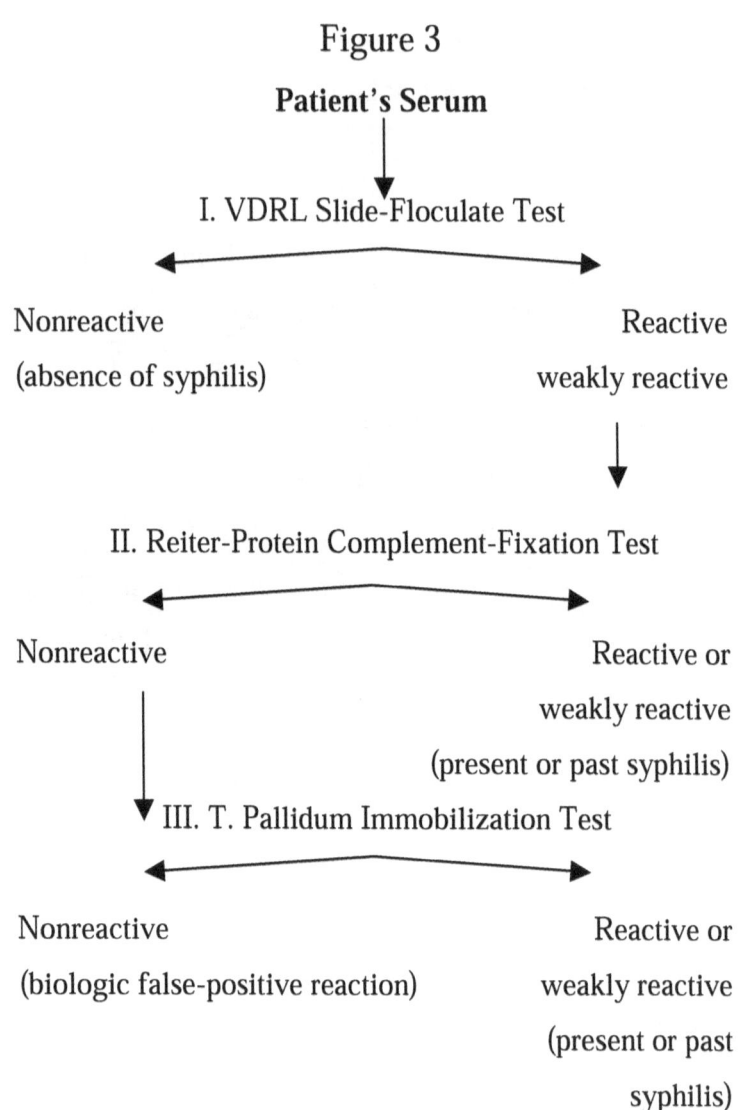

Patient's Serum

I. VDRL Slide-Floculate Test

Nonreactive
(absence of syphilis)

Reactive
weakly reactive

II. Reiter-Protein Complement-Fixation Test

Nonreactive

Reactive or
weakly reactive
(present or past syphilis)

III. T. Pallidum Immobilization Test

Nonreactive
(biologic false-positive reaction)

Reactive or
weakly reactive
(present or past
syphilis)

This flow chart can be represented in a signal flow diagram as represented in Figure 4.

Figure 4

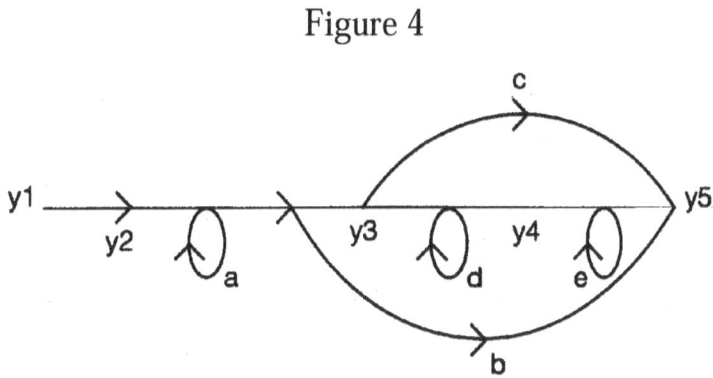

The following is a table, Table 3, representing the items in Figure 3.

Table 3

y1 is patient's serum and is the signal source.

y2 is the VDRL Test.

a is the nonreactive response and is a feedback
loop.

b is reactive and is a path.

y3 is the Reiter-Protein Complement Test.

c is reactive and is a path.

d is nonreactive and is a feedback loop.

y4 is T. Pallidum Test.

e is nonreactive and is a feedback loop.

y5 is reactive and is the sink.

The path gain is the line between y1 and y5 and

y2, y3 and y4 are the test variables[4] .

Model C is the possible effects on laboratory tests using cyclosporine and is represented by the signal flow diagram in Figure 5.

Figure 5

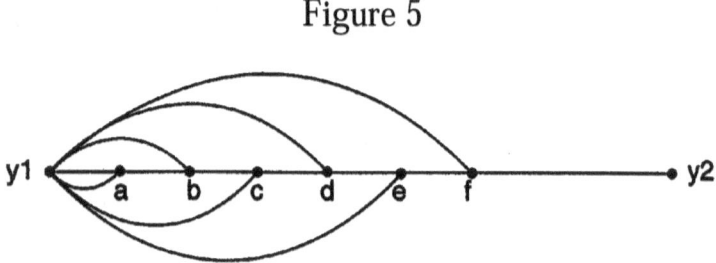

The following table, Table 4, represents the items in Figure 5.

---

[4] For more detail on Serologic Testing of syphilis see Appendix D.

Table 4

y1 is the source.

a is the blood cell count (decrease).

b is the blood potassium level (increase).

c is the blood uric acid level (increase).

d is the blood platelets, white cells, and magnesium (decrease).

e is the liver function tests (increase).

f is the kidney function tests (increase).

y2 is the sink.

All of these conditions represent negative reactions to the laboratory tests[5] .

---

[5] For more information on the drug cyclosporine see Appendix E.

<u>Discussion and Analysis of Results</u>

In reviewing the data from the three models of signal flow diagrams in the <u>Pharmacological Model</u> in Chapter 6 of this dissertation, the difference between block diagrams and signal flow diagrams becomes strikingly apparent. In Model A the signal flow diagram graphs the rank by percentile of reactions to the six antigens of the hypersensitivity skin test. In this model the contrast is not so much between a table or chart of information but rather how such information would look in a signal flow diagram.

Upon first inspection of the two systems of representing information it is apparent that the signal flow diagram is providing a specific type of information more quickly and more clearly than the table of information, mainly that the rank of percentile is more obvious by the use of branches and nodes on the signal flow diagram than is numerically or alphabetically represented in the table. This is a classic case of semantical verses visual information as the table is a linguistic and numerical device and the signal flow diagram is a visual or graphic device.

In taking into account the space needed to generate the amount of information, the signal flow diagram is superior in that it takes less space, in this case about one quarter of the space of the table, and has a clear directional sense, pointing arrows, and a corresponding hierarchy of branches and nodes representing the individual items and there ranking by percent. In Model B the Triple-Test Plan for Serologic Diagnosis of Syphilis is first represented by a flow chart and then by a signal flow diagram. The flow chart is a popular method of representing a process hierarchy and is found in most information oriented disciplines.

The flow chart is a visually clear representation of the information and affords more information than a table or a chart but as compared to the signal flow diagram, it is again overly complex and time consuming when matched with the simplicity of the signal flow diagram. Notice that the nodes denoting the nonreactive response are clearly represented by loops and that the branches denoting a reactive response are all branched into the reactive sink node of the diagram. This is a more clear representation of the information than can be inferred from the flow chart and takes up less space than the flow chart even with the corresponding table of data.

Model C is the possible effects on laboratory tests using cyclosporine and is a diagram from data collected from a clinical case study of the drug. The signal flow diagram is more effective in describing and detailing the drug effects than the raw data and has the added benefit of clearly representing increases and decreases by the direction of the branch arrows. It is clear from these examples that signal flow diagrams are superior to tables, charts, flow charts, numerical and word representations, but it is also clear that signal flow diagrams are superior to other graphical representations most notably the formal or block diagram.

In the chapter <u>Signal Flow Verses Block Diagrams</u> in this dissertation, the complexity, time and space involved to represent equations in block diagrams becomes apparent as the more complex a system, the more involved becomes the block diagram until it becomes a daunting task not only to design the block diagram but to read and comprehend the information represented with the diagrams. At a certain saturation point the block diagram becomes redundant as the information is no longer being imparted by this method of graphic representation. The block diagram reaches this saturation point long before the signal flow diagram does and is a test of the simplicity of the design of such a graphic representation.

From the point of view of the information sciences the signal flow diagram is a superior method of visually displaying mathematical equations into simple but accurate representations of information. Cybernetics is involved on one level in that the integration of information from raw data to usable information with complexity, time and space being contraints to how the information is processed, clearly demonstrates the importance of simple communication systems such as the signal flow diagrams in increasing the efficiency and feedback response, on all levels, of the information sciences.

This is clearly a communication science as it is an information science in that mathematical, numerical and word models are represented by this form of visual graphic display and that the signal flow diagram is the superior form of this representation.

<u>Conclusions</u>

The use of signal flow diagrams in graphing feedback data for pharmacology has been shown to be of great importance in making the process of drug diagnosis more efficient, effective and more accurate than either block diagrams or equations in processing complex systems. This method can substantially reduce staff burnout, costs and needless errors that will have substantial positive effects on patient costs, well being and success rates.

The following can be said of the strengths of the signal flow diagram.

1. The signal flow diagram is a simplified graphic representation of mathematical, numerical and word models and that these models are best expressed by the signal flow diagram.

2. The nodes and branches of the signal flow diagram can be a symbol of all types of data and information and the branches' arrows can represent loops, increases and decreases in relation to the information being

represented and the nodes denoting a hierarchy of the information being represented.

3. Time and space are saved by the use of signal flow diagrams and this can be important when labor, cost and efficiency factors are involved.

4. The complexity of the information is made more simple and more visually clear by the use of signal flow diagrams and that this simplicity is inherent in such a graphical method.

5. The signal flow diagram is superior to formal or block diagrams as block diagrams are inherently weak in the area of simplification and ease of use and are also time and space sensitive.

6. Both equations and raw data are inferior to signal flow diagrams in that equations are long, complex and time consuming and raw data is marginal at imparting specific information when compared to the signal flow diagram.

7. The signal flow diagram is superior to tables, charts and flow charts in that it more readily accepts large quantities of information and represents them in the

most accurate and simplistic manner, something that tables, charts and flow charts perform with limited success.

8. The information saturation point of signal flow diagrams is higher than other forms of information representation.

9. Overall simplicity of conception, use and understanding is the main point of interest and support for the signal flow diagram.

10. Accuracy of the signal flow diagram is in the simplicity of its use.

The signal flow diagram is the graphical method of choice for the representation of mathematical, numerical and word models and is superior to equations, raw data, tables, charts, flow charts, and block diagrams in representing the desired information.

Recommendations For Future Work

It is hoped that this study will inspire a more comprehensive program in the future and is a strong indicator of the successful application and use of signal flow diagrams in pharmacology.

An interesting development of the signal flow diagram would be the color coding of the branches and nodes of the signal flow diagram to give greater representation and contrast to each of the branches and nodes. Another development would be the use of computer graphics to represent such diagrams and ease the making of such diagrams by having them integrated into the software programs.

A greater variety of data and information should be used in the future to give a wider representation of the uses of signal flow diagrams and the testing of other graphic and display oriented visual information. Such future research can only strengthen the use of such visual information for a wider use in all the information related disciplines.

Appendix A

The following is an unpublished chemistry paper that was done as Director of Research at Advanced Human Design in Cupertino, California U.S.A. and is the bases for this dissertation.

The Use of Signal Flow Diagrams in Pharmacology

by Bradley S. Tice, Director
Advanced Human Design

The use of signal flow diagrams are common in fields such as engineering. A practical use can be made in the field of pharmacology for similar reasons-ease of use. These graphs can be used to solve complex linear, multiloop systems in less time than either block diagrams or equations (Macmillan, Higgins, and Naslin, 1964:4). A signal flow graph is the topological representation of a set of linear equations as represented by the following equation

$$\text{Equation 1:} \quad y_i = \sum_j^n a_{ij} x_j, \quad i = 1, \ldots, n$$

Branches and nodes are used to represent a set of equations in a signal flow graph. Each node represents a variable in the system, like node i represents variable y in equation 1. Branches represent the different variables such as branch ij relates variable yi to yj where the branch originates at node i and terminates at node j in equation 1 (Shinners, 1964:25). The following set of linear equations are represented in the signal flow graph in Figure 1 (Shinners, 1964:25).

$$y2 = \quad ay, \quad +by2 \qquad\qquad +cy4$$
$$y3 = \qquad\qquad dy2$$
$$y4 = \quad ey1 \qquad\quad +fy3$$
$$y5 = \qquad\qquad\qquad +gy3 \quad +hy4$$

Figure 1

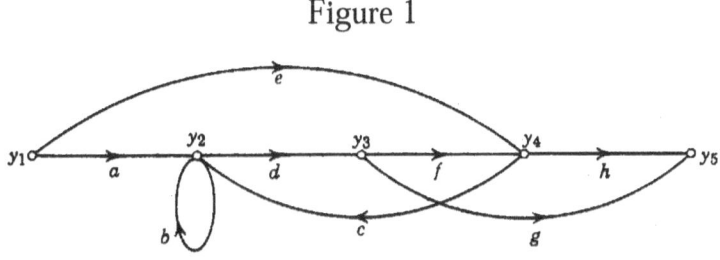

By using a signal flow diagram to represent time of drug delivery, behavioral response to the drug, and drug dosage, the cumulative sequence of events is visually represented for ease of interpretation of data.

This type of graphic representation of complex linear equations makes interpretation of drug analysis that much more efficient and effective for evaluation.

References

Shinners, S.M. (1964). Control System Design. New York: John Wiley & Sons, Inc.

Macmillan, R.H., Higgins, T.J., and Naslin, P.(eds). (1964). Progress in Control Engineering. New York: Academic Press Inc., Publishers.

Appendix B

The following is an unpublished chemistry article that was done as Director of Research for Advanced Human Design in Cupertino California U.S.A.

Signal Flow Diagrams verses Block Diagrams:

The Simpler Road.

by Bradley S. Tice, Director

Advanced Human Design

The strongest points in using a signal flow diagram over block diagrams is that they are more general in use and are a more pronounced rationalization afforded by diagrammatic simplification (Macmillian, Higgins, and Naslin, 1964:4).

Methods

A block diagram is a graphical representation of interconnected elements which form a system that differ only in their dynamic properties (Solodov, 1966:44). A signal flow diagram represents a set of equations by means of branches and nodes. The use of the signal flow diagram permits the solution practically by visual inspection (Shinners' 1964:25). A model of a block diagram will be compared to a model of a signal flow diagram for a comparative and contrastive analysis of the two graphic systems.

Examples

The following is a model of a standard single-loop-feedback system using a block diagram in Figure 1(Eveleigh, 1960:61).

Figure 1

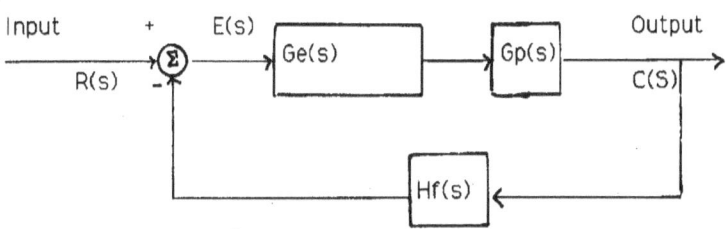

The next model is a signal flow diagram for the following algebraic equation system, Table 1, and represented by a signal flow diagram in Figure 2 (Macmillian, 1964: 5).

Table 1

x = x

x = t x + t x

x = t x + t x +t x

Figure 2

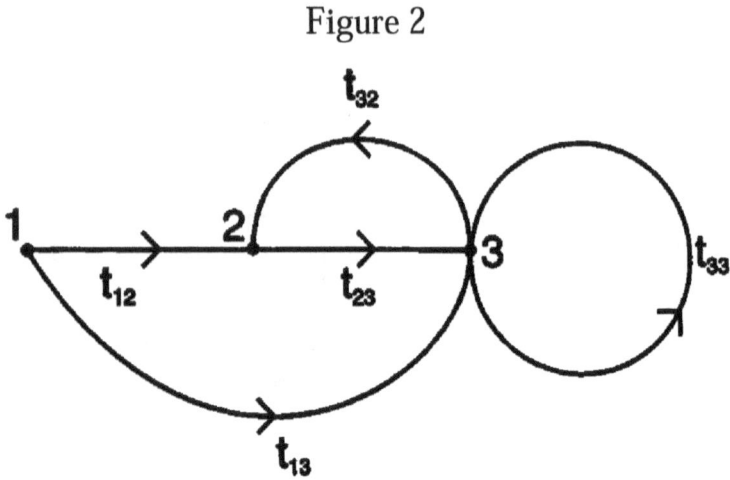

Results

In examining the block diagram in Figure 1 and the signal flow diagram in Figure 2 a comparison between the two systems can be made on a point by point analysis.

1. The signal flow diagram is a simplified graphic representation of mathematical, numerical and word models and that these models are best expressed by the signal flow diagram.

2. The nodes and branches of the signal flow diagram can be a symbol of all types of data and information and the branches' arrows can represent loops, increases and decreases in relation to the information being represented and the nodes denoting a hierarchy of the information being represented.

3. Time and space are saved by the use of signal flow diagrams and this can be important when labor, cost and efficiency factors are involved.

4. The complexity of the information is made more simple and more visually clear by the use of signal flow diagrams and that this simplicity is inherent in such a graphical method.

5. The signal flow diagram is superior to formal or block diagrams as block diagrams are inherently weak in the area of simplification and easy of use and are also time and space sensitive.

6. Both equations and raw data are inferior to signal flow diagrams in that equations are long, complex and time consuming and raw data is marginal at imparting

specific information when compared to the signal flow diagram.

7. The signal flow diagram is superior to tables, charts and flow charts in that it more readily accepts large quantities of information and represents them in the most accurate and simplistic manner, something that tables, charts and flow charts perform with limited success.

8. The information saturation point of signal flow diagrams is higher than other forms of information representation.

9. Overall simplicity of conception, use and understanding is the main point of interest and support for the signal flow diagram.

10. Accuracy of the signal flow diagram is in the simplicity of it's use.

Discussion

The strengths of the signal flow diagram over the block diagram are very pronounced on all levels of evaluation and merit a wider application of such a graphing system to other information display categories. These qualities of simplicity, generality of use, and clarity of realization makes the signal flow diagram a more efficient method than block diagrams on solving a systems problem.

References

Eveleigh, V.W. (1960). <u>Adaptive Control and Optimization Techniques</u>. New York: McGraw-Hill Book Company.

Macmillian, R.H., Higgins, T.J., and Naslin, P. (eds). (1964). <u>Progress in Control Engineering</u>. New York: Academic Press Inc., Publishers

Shinners, S.M. (1964). <u>Control System Design</u>. New York: John Wiley & Sons, Inc.

Solodov, A.V. (1966). <u>Linear Automatic Control Systems with Varying Parameters</u>. New York: American Elsevier Publishing Company, Inc.

Appendix C

The data and information in Model A is taken from Lynn E. Spiter's "Delayed Hypersensitivity Skin Testing" from Rose and Friedman's <u>Manual of Clinical Immunology</u> (Washington: American Society for Microbiology, 1980).

Skin testing is the most important clinical assessment of the status of cellular immune responses in patients (Rose and Friedman, 1980: 200). In essence, a delayed hypersensitivity reaction is if a red bump develops at the site of injection of a test antigen, indicating that the afferent, central and efferent limbs of the immune system response is intact and that the patient's ability to start a nonspecific inflammatory response is intact (Rose and Friedman, 1980: 200). There are four clinical indications that skin tests are used to assess.

1. To assess whether there is diminished delayed hypersensitivity or anergy in selected patients.
2. To asses the results of immunotherapy.
3. To follow the course of the disease process.
4. As an aid to diagnosis infectious diseases.

Procedures

The test procedures are as follows.

1. Battery of six skin test antigens.
2. Repeat the test in higher antigen concentrations when the tests are negative with the intermediate strength.
3. Observe and record the results in millimeters of erythema and induration at 24 hour and 48 hour intervals (Rose and Friedman, 1980: 202).

The six antigens are to be injected into a marked region, indelible pencil, on the patient, usually the forearms, in a subcutaneous fashion leaving a 5 cm diameter bleb. Recording the results after 24 hours (Rose and Friedman, 1980: 202).

Interpretation

A positive response is determined when a 5mm or more of induration is found at the test site 48 hours after the injection of the test antigen.

Antigens

The six antigens are as follows.

1. Candidin

2. Mixed respiratory vaccine

3. PPD

4. SK-SD

5. Staphage Lysate

6. Trichoplytin

Note: Both Coccidioidin and Mumps were used in the testing as they are also clinical indicators of possible anergy.

Appendix D

The data and information for the "Triple-Test Plan for Serologic Diagnosis of Syphilis" is from Levinson's and MacFate's <u>Clinical Laboratory Diagnosis</u> (Philadelphia: Lea & Febiger, 1969).

There are many serologic tests that can be used to detect the etiologic agent for syphilis, the Treponema pallidum, and its antibodies, in serums and cerebrospinal fluids. The tests maybe treponemal or nontreponemal (Levinson and MacFate, 1969: 671). Nontreponemal tests include precipitation and flocculation methods, such as VDRL (Veneral Disease Research Laboratory) test and the Kahn test, and those complement-fixation tests which use antigens derived from animal tissue extracts such as the Kolmer test (Levison and MacFate, 1969: 671).

Treponemal tests include agglutination and treponemal immobilization methods, such as the TPI (Treponema Pallidum Immobilization) test, and those complement-fixation tests which use antigens extracted from either virulent or saprophytic strains of Treponema pallidum, such as the Reiter Protein Complement-Fixation test (Levison and MacFate, 1969: 671). Most positive serologic tests obtained with nontreponemal antigens are due to syphilis (Levisnson and MacFate, 1969: 671).

A biologic false-positive reaction may be due to the presence of antibody like substances similar to the antibodies produced in syphilis. To overcome these false reactions, tests were developed using the Treponema pallidum itself or an antigen derived from the organism. These tests are specific and biologically false-positive reactors give negative results (Levinson and MacFate, 1969: 671)

Appendix E

The data and information for cyclosporine is taken from Drug Facts and Comparisons (St. Louis: Facts and Comparisons, 1995), Harkness's Drug Interactions Guidebook (Englewood Cliffs: Prentice Hall, 1991), and Rybacki and Long's The Essential Guide to Prescription Drugs (New York: Harper Perennial, 1996).

Cyclosporine is a cyclic polypeptide immunosuppressant. The avaliable dosage forms and strength are as follows.

1. Capsules, soft gelatin-25mg, 100mg.
2. Injection, intravenous-50mg per ml.
3. Oral solution-100mg per ml.

Usual dosage range is 15mg per day taken 4 to 12 hours before transplant surgery. Reduced to 5 to 10mg per day two weeks after surgery (Rybacki and Long, 1996: 266).

Cyclosporine blood levels should be closely monitored along with the serum creatinine. Regulate, ie. lower dosage level, dose of cyclosporine as needed (Harkness, 1991: 85).

Distribution of cyclosporine in the body is as follows. 33% to 47% is in plasma.

4% to 9% in lymphocytes.
5% to 12% in granulocytes.
41% to 58% in erythrocytes.

Note: Blood level monitoring of cyclosporine may be useful in patient management. While no fixed relationships have been established, 24 hour trough values of 250 to 800 ng/ml (whole blood, RIA) or 50 to 300 ng/ml (plasma, RIA) appear to minimize side effects and rejection events (Facts and Comparisons, 1995: 3108).

References

1. Cronin, K.M., Lane, G.H., and Peirce, A.G. (1983). Flow Chartes-Clinical Decision Making in Nursing. Philadelphia: J.B. Lippincott.

2. Chang, S.L..(1961). Synthesis of Optimum Control Systems. New York: McGraw-Hill Book Company, Inc.

3. Porter, A. (1950). Introduction to Servomechanisms. London: Methuen & Company LTD.

4. Murrill, P.W. (1967). Automatic Control of Processes. Scranton: International Textbook Company.

5. Koenig, H.E. and Blackwell, W.A. (1961). Electromechanical System Theory. New York: McGraw-Hill Book Company, Inc.

6. Newton, G.C. and Gould, L.A. (1957). <u>Analytical Design of Linear Feedback Controls.</u> New York: John Wiley & Sons, Inc.

7. Smith, O.J.M. (1958). <u>Feedback Control Systems</u>. New York: McGraw-Hill Book Company, Inc.

8. Popov, E.P. (1962). <u>The Dynamics of Automatic Control Systems</u>. London: Pergamon Press LTD.

9. James, H.M., Nicholes, N.B. and Phillips, R.S. (1964). <u>The Theory of Servomechanisms</u>. Lexington: Boston Technical Publishers, Inc.

10. Holzbock, W.G. (1958). <u>Automatic Control</u>. New York: Reinhold Publish Corporation.

11. Zeines, B. (1959). <u>Servomechanism Fundamentals</u>. New York: McGraw-Hill Book Company, Inc.

12. Horowitz, I.M. (1963). <u>Synthesis of Feedback Systems</u>. New York: Academic Press.

13. Ku, Y.H. (1962). <u>Analysis and Control of Linear Systems</u>. Scranton: International Textbook Company.

14. Hardie, A.M. (1964). <u>The Elements of Feedback and Control</u>. New York: Oxford University Press.

15. Oldenbourg, R.C. and Sartorius, H. (1953)."A Uniform Approach to the Optimum Adjustment of Control Loops" in <u>Frequency Response Symposium</u>. New York: The American Society for Mechanical Engineers.

16. Kipiniak, W. (1961). <u>Dynamic Optimization and Control</u>. New York: The MIT Press and John Wiley & Sons, Inc.

17. Hadley, W.A. and Longobardo, G. (1963). <u>Automatic Process Control</u>. Palo Alto: Addison-Wesley Publishing Company, Inc.

18. Chubb, B.A. (1967). <u>Modern Analytical Design of Instrument Servomechanisms</u>. Palo Alto: Addison-Wesley Publishing Company.

19. Gille, J.C., Delegrin, M.J., and Deculine, P. (1959). <u>Feedback Control Systems</u>. New York: McGraw-Hill Book Company, Inc.

20. Bulliet, L. J. (1967). <u>Servomechanisms</u>. Menlo Park: Addison-Wesley Publishing Company.

21. Bulgakov, A.A. (1965). <u>Energetic Processes in Follow Up Electrical Control Systems</u>. New York: The Macmillan Company.

22. Tucker, G.K. and Willis, D.M. (1958). <u>A Simplified Technique of Control System Engineering</u>. Philadelphia: Honeywell Regulatory Company.

23. Rusinoff, S.E. (1957). <u>Automation in Practice</u>. Chicago: American Technical Society.

24. Seifert, W.W. and Steeg, C.W. (1960). <u>Control Systems Engineering</u>. New York: McGraw-Hill Book Company, Inc.

25. Derusso, P.M., Roy, R.J. and Close, C.M. (1965). <u>State Variables for Engineers</u>. New York: John Wiley & Sons.

26. Ahrendt, W.R. (1954). <u>Servomechanism Practice</u>. New York: McGraw-Hill Book Company, Inc.

27. Bowen, J.L. and Schultheises, P.M. (1961). <u>Introduction to the Design of Servomechanisms</u>. New York: John Wiley and Sons, Inc.

28. Nagel, E. (1948) <u>Automatic Control</u>. New York: Simon and Schuster.

29. Warren, J.E. (1967). Control Instrument Mechanisms. Kansas City: The Bobbs-Merrill Company, Inc.

30. West, J.C. (1953). Textbook of Servomechanisms. New York: The Macmillan Company.

31. Qvarnstrom, B., Schutt, T., and Runnstrom-Reio, V. (1965). Instruments and Measurements. New York: Academic Press.

32. Oguztoreli, M.N. (1966). Time-Lag Control Systems. New York: Academic Press.

33. Thaler, G.J. and Pastel, M.P. (1962). Analysis and Design of Nonlinear Feedback Control Systems. New York: McGraw-Hill Book Company, Inc.

34. Peschon, J. (1965). Disciplines and Techniques of Systems Control. New York: Blaisdell Publishing Company.

35. Tsien, H.S. (1954). <u>Engineering Cybernetics</u>. New York: McGraw-Hill Book Company, Inc.

36. Tou, J.T. (1964). <u>Modern Control Theory</u>. San Francisco: McGraw-Hill Book Company.

37. Kuo, B.C. (1962). <u>Automatic Control Systems</u>. Englewood Cliffs: Prentice-Hall, Inc.

38. Solodovnikov, V.V. (1965). <u>Statistical Dynamics of Linear Automatic Control Systems</u>. London: D. Van Nostrand Company, LTD.

39. Wilts, C.H. (1960). <u>Principles of Feedback Control</u>. London: Addison-Wesley Publishing Company, Inc.

40. Mishkin, E. and Braun, L. (1961). <u>Adaptive Control Systems</u>. New York: McGraw-Hill Book Company, Inc.

41. Tomovic, R. (1963). <u>Sensitivity Analysis of Dynamic Systems</u>. New York: McGraw-Hill Book Company, Inc.

42. Lindoff, D.P. (1965). <u>Theory of Sampled-Data Control System</u>. New York: John Wiley & Sons, Inc.

43. Wiener, N. (1948). <u>Cybernetics: or Control and Communication in the Animal and the Machine</u>. Cambridge: The MIT Press.

44. Wiener, N. (1950). <u>Human Use of Human Beings: Cybernetics and Society</u>. New York: Avon Books.

45. Cannon, W. (1932). <u>The Wisdom of the Body</u>. New York: W.W. Norton & Company, Inc.

46. Adolph, E.F. (1960). <u>The Development of Homeostasis.</u> London: Academic Press.

47. Langley, L.L. (1965). <u>Homeostasis</u>.

New York: Reinhold Publishing Corporation.

48. Henderson, L.J. (1913). <u>The Fitness of the Environment</u>. Boston: Beacon Press.

49. Lwoff, A. (1962). <u>Biological Order</u>. Cambridge: The MIT Press.

50. Shannon, C. E. and Weaver, W. (1949). <u>The Mathematical Theory of Communication</u>. Urbana: University of Illinois Press.

51. Tufte, E. (1983). <u>The Visual Display of Quantitative Information</u>. Cheshire: Graphics Press.

52. Tufte, E. (1990). <u>Envisioning Information</u>. Cheshire: Graphics Press.

53. Thaler, G.J. (1955). <u>Elements of Servomechanism Theory</u>. New York: McGraw-Hill Book Company, Inc.

54. Izawa, K. (1963). Introduction to Automatic Control. Amsterdam: Elsevier Publishing Company.

55. (1958). Principles of Frequency Response. Pittsburgh: Instrument Society of America.

56. Doebelin, E.O. (1962). Dynamic Analysis and Feedback Control. New York: McGraw-Hill Book Company, Inc.

57. Del Toro, V, and Parker, S.R. (1960). Principles of Control Systems Engineering. New York: McGraw-Hill Book Company, Inc.

58. Solodov, A.V. and Fuller, A.T. (1966). Linear Automatic Control Systems with Varying Parameters. New York: American Elsevier Publishing Company, Inc.

59. Macmillan, R.H. (1964). Progress in Control Engineering. New York: Academic Press.

60. West, J.C. (1960). <u>Analytical Techniques for Non-Linear Control Systems.</u> New York: D. Van Nostrand Company, Inc.

61. Caruthers, F.P. and Levenstein, H. (1963). <u>Adaptive Control Systems</u>. New York: A Pergamon Press Book.

62. Bellazzi, R. (1993) "Drug delivery optimization through Bayesian networks; an application to erythropoietin therapy in uremic anemia." in Computers and Biomedical Research, 1993 June 26(3): 274-93.

63. Plouffe, L (1992). "Biological modeling on a microcomputer using standard spreadsheet and equation solver programs: the hypothalamic-pituitary-ovarian axis as an example." In Computers and Biomedical Research. 1992 April 25(2): 117-30.

64. Lenert, L.A., Lurie, J., Sheiner, L.B., Coleman, R., Klostermann, H., and Blaschke, T.F. (1992).

"Advanced computer programs for drug dosing that combine pharmacokinetic and symbolic modeling of patients." in Computers and Biomedical Research. 1992 February 25(1): 29-42.

65. Lago, P.J. (1992). "Open-loop stochastic control of pharmacokinetic systems: a new method for design of dosing regimens." in Computers and Biomedical Research. 1992 February 25(1): 85-100.

66. Peslin, R., Saunier, C., Duvivier, C. and Marchand, M. (1995). "Analysis of low-frequency lung impedance in rabbits with nonlinear models." in the Journal of Applied Physiology. September 1995 79(3): 771-80.

67. DeVries, S.H. and Baylor, D.A. (1995). An alternative pathway for signal flow from rod photoreceptors to ganglion cells in mammalian retina." in the Proceedings of the National Academy of Sciences. November 1, 1995 92(23): 10658-62.

68. Wei, D., Shea, M., Saidel, G.M. and Jones, S.C. (1993). "Validation of continuous thermal measurement of cerebral blood flow by arterial pressure change." in Journal of Cerebral Blood Flow and Metabolism. July 1993 13(4): 693-701.

69. Panteleev, Q.A., Vaneeva, L.I., Demchenko, E.N., and Semenova, G.V. (1987). "Problems of selecting the reagent concentration in the solid-phase immunoenzyme method for determining antibody concentrations." in Zhurnal Mikrobiologii Epidemiologii i Immunobiologii. April 1987 (4): 80-5.

70. Clark, E.M. (1980). "A new design for an audio dosimeter." In the American Industrial Hygiene Association Journal. October 1980 41(10): 700-3.

71. Hubener, K.H. (1978). "Conceptions for optimization in radiation therapy." in Strahlentherapie. December 1978 154(12): 858-60.

72. Dubourg, O., Chikli, F., and Delorme, G. (1995). "Contrast echocardiography in clinical practice." in the Journal d'Echographie et de Medecine par Ultrasons. 1995 16/4:143-152.

73. Fukuyama, T., Yamaoka, K., Ohata, Y., and Nakagawa, T. (1994). "A new analysis method for disposition kinetics of enterohepatic circulation of diclofenac in rats." in Drug Metabolism Dispositions. 1994 22/3: 479-485.

74. Monzon, J.E. and Guillen, S.G. (1985). "Current defibrillator: New instrument of programmed current for research and clinical use." in IEEE Transactions Bio-Medical Engineering. 1985 32/11: 928-34.

75. Guenther, B., Morgado, E. and Penna, M. (1974). "Equifinality on the circulatory system of some mammals." in Pflug Archiv European Journal of Physiology. 1974 348/4: 343-352.

76. Krinsky, J.A. (1988). "Testing fiber optic data links for sensitivity to high gamma radiation dose rates." Delivered as a paper at the Second Annual Military Fiber Optics Conference-West in 1988 and publishing in Boston: Information Gatekeepers.

77. Ladu, M. and Randaccio, P. (1980). "A simple device for photographic dosimeter calibration." in Fisica e Tecnologia. 3/4: 276-80.

78. Coco, C. (1979). "A more simple PROM programmer." In Elettronica Oggi February 1979 2: 109-10, 112, 114-15.

79. Feil, H.J. (1979). "Measuring, controlling and dose-rating with miniture oval-gear counters." in Und-oder-Nor & Steuerungstechnik. 1979 1-2: 32-3.

80. Heister, H. (1979). "New developments in x-ray television." In Fernseh-und Kino-Technik. February 1979 33/2: 41-2.

81. Ranjbaran, E. "Modeling of Electrical Engineers." dissertation. University of Missouri, Columbia 1974.

82. Menshutkin, V.V, Suvorova, T.P. and Balonov, L.Y. (1981). "Models of functional shutdowns of 1 hemisphere and neuro pharmacological effects on deep cerebral structures." In Fiziologiya Cheloveka. 7/5: 880-888.

83. Danilov, V.A, (1984). "Television microscopes for studying biological microscopic material." in Biomedical Engineering. July-August 1984 18/4: 125-130.

84. Ranjbaran, S.E. (1980). "State variable techniques in pharmacokinetics using computer graphics." in IEEE Engineering Medical and Biological Society Annual Conference September 28-30, 1980 in Washington D.C.: IEEE.

85. Shinners, S.M. (1964). <u>Control System Design</u>. New York: John Wiley & Sons, Inc.

86. Macmillian, R.H., Higgins, T.J. and Naslin, P. (1964). <u>Progress in Control Engineering.</u> New York: Academic Press Inc., Publishers.

87. Solodov, A.V. (1966). <u>Linear Automatic Control Systems with Varying Parameters</u>. New York: American Elsevier Publishing Company, Inc.

88. Facts and Comparisons. (1995). <u>Drug Facts and Comparisons</u>. St. Louis: Facts and Comparisons.

89. Harkness, R. (1991). <u>Drug Interactions Guidebook</u>. Englewood Cliffs: Prentice Hall.

90. Levinson, S.A. and MacFate, R.P. (1969). <u>Clinical Laboratory Diagnosis</u>. Philadelphia: Lea & Febiger.

91. Rose, N.R. and Friedman, H. (1980). Manual of Clinical Immunology. Washington: American Society for Microbiology.

92. Rybacki, J.J and Long, J.W. (1996). The Essential Guide to Prescription Drugs. New York: Harper Perennial.

About the Author

Dr. Tice is Director and Institute Professor of Chemistry at Advanced Human Design located in Cupertino, California U.S.A. Dr. Tice is, or has been, a member of the following organizations: American Chemical Society, American Society for Microbiology, Association for Computing Machinery, I. E. E. E., The American Physical Society, American Institute of Aeronautics and Astronautics, the Committee on Space Research (COSPAR), and a founding member of The Mars Society. Dr. Tice is listed in the 53rd and 54th editions of Marquis Who's Who in America and the 16th and 17th editions of Marquis Who's Who in the World. Dr. Tice is also CEO of Tice Pharmaceuticals that is located in San Jose, California U.S.A.